LITTLE BOOK OF SCOTTISH BEASTIES

THE LITTLE BOOK OF

SCOTTISH BEASTIES

TIM KIRBY

First published in 2019 by
Birlinn Limited
West Newington House
10 Newington Road
Edinburgh
EH9 1QS

www.birlinn.co.uk

ISBN: 978 1 78027 627 4

British Library Cataloguing-in-Publication Data
A catalogue record for this book
is available from the British Library

Designed and typeset by Mark Blackadder

Printed and bound by Gutenberg Press, Malta

INTRODUCTION

The relationship between people and animals goes back tens of thousands of years, and animal names form a rich part of any language's vocabulary. In Latin they are called *animalia* because they are animated by life. The Scots call them 'beasties', an affectionate term that encompasses creatures of all shapes and sizes.

During medieval times books called bestiaries became popular. These contain information about real and imaginary animals and are illustrated with beautiful illuminated artwork. In its own way this book is a continuation of that tradition, containing as it does a mixture of myth and legend together with more contemporary observations about Scotland's remarkable wildlife.

ARBROATH SMOKIE

The Arbroath Smokie is a fish so called because of the smoky flavour of its flesh. It is most commonly seen in Arbroath living among potatoes, but is also said to be found in the Barry Burn around the golf links of Carnoustie, where it is often hunted by men with sticks, without much success.

EWAN'S
LAGER
32Re

BADGER

The badger (or *brochlach*, or *broc*) lives underground in holes it makes by digging with its feet. It is said that to carry the soil out, a badger lies on its back with a stick in its mouth. The soil is then piled on its belly and the other badgers pull it out by the stick. The badger is an avid follower of football and so has a stadium, Ibrox, meaning 'home of the badgers', named after it.

BAGPIPE

The bagpipe is known for its very loud cry, which it makes when stressed. This attribute is commonly provoked by humans as a form of music. The making of instruments from animals is known in other parts of the world, but Scotland is unique in that the bagpipe is 'played' while it is still alive.

OCH, NO

BEAVER

The beaver was hunted for the purported medicinal properties of its testicles. Being a clever animal, it learned to survive by biting them off and showing the castrated area to its pursuer. Since the beaver's reintroduction to Scotland, such displays have been replaced by a less painful, and thankfully infrequent, human practice.

MORNING

BEITHIR

The *beithir* is the largest and most deadly of serpents. A supernatural creature, it is created when a decapitated serpent rejoins its body. The old way of preventing a *beithir* from arising was to make sure that the head was removed to some distance. Current thinking is that they are probably best left undisturbed.

ARGH

BOOBRIE

The boobrie is a great, ravenous sea bird that frequents the watercourses of Argyll. It feeds upon cattle and sheep, and its call is so terrifying it is said to have once driven a minister from his house.

YOU ARE SUCH
AN IDIOT

CAPERCAILLIE

The name capercaillie means 'great cock of the woods'. The male is seen mostly dancing and shouting and performing japes and capers, while the female would rather they did something more useful.

CEASG

The *ceasg* is a type of mermaid, half girl, half salmon. She is kindly to sailors and will protect them from storms. It has been known for marriages to occur between *ceasg* and men, usually fishermen, who are not perturbed by the fishy aroma.

BURP

CIREIN-CRÒIN

The *cirein-cròin* is a fearsome shape-shifting sea beast. It is often so large it can feed on seven whales, followed by cranachan for dessert, but it can also become as small as a fish, so as to better lure its prey.

OH, COME ALONG NOW, HAMISH, LEAVE THE POOR WEE MAN ALONE

CÙ-SÌTH

The *cù-sìth* is a fairy dog of the devil's making. To see one is often taken to be a sign of death, but it may be that the observer is on a popular dog-walking route.

DINOSAUR

The animal called the dinosaur is an ancient great desert lizard from before the time of Scottish rain. It has names like McSaltopus, McSauropod, McCeratops and Dougie. It has been a long time since one has been seen, but footprints are occasionally found.

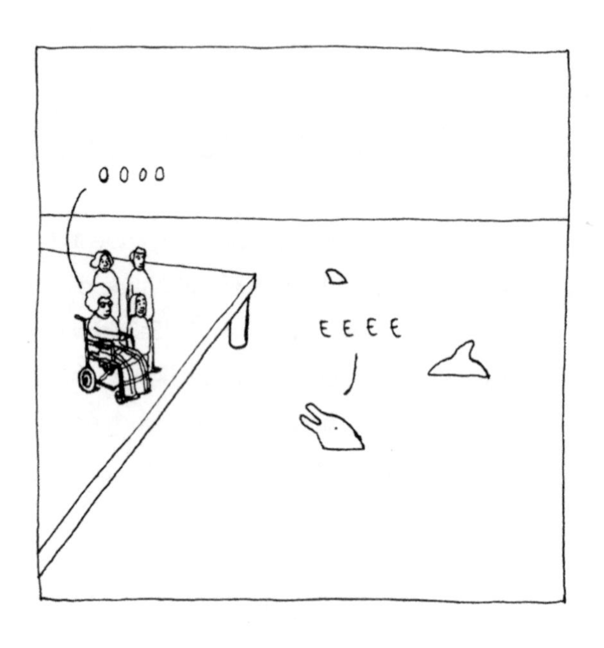

OOOO
EEEE

DOLPHIN

The dolphin is a water-dwelling animal famed for acrobatics and funny noises. They are most commonly found off the east coast of Scotland, around Moray. Like those other producers of funny noises – grannies and favourite aunts – people are always pleased to see them and will make special trips for the purpose.

FOX
&
HAGGIS
REAL
ALES
AYE, IT'S THE DEVIL'S DRINK ALRIGHT

FOX

It is often said that the fox never walks in a straight line and that this is because it is a devious animal, but it may just be due to the build-up of McEwan's ale in the water table. Like beer, the fox is said to be a symbol of the devil.

HEY,
COME BACK
WITH MY
TEA CAKE

GOLDEN EAGLE

The golden eagle is a winged, feathered beast. It gets its name because it is eagle-eyed and can spot a Tunnock's Tea Cake from three kilometres. It is said that if its sight diminishes it flies up to the sun and its vision is restored.

CAN I GET HIS AUTOGRAPH FIRST?

GROUSE

The grouse is a type of bird that lives mostly among the heather of the Scottish moorlands, hiding much of the year either from fame and fortune or from the tweed-haired creatures that pursue it between the 'glorious twelfth' of August and the tenth of December.

HE MUST BE
AN ANTI-CLOCKWISE
ONE

HAGGIS

Legend has it that the haggis lives on mountain-sides, and that of its four legs two are shorter on one side than the other to facilitate staying upright on the slopes. It is said that it is the great chieftain of the pudding race. Confirmed wild sightings are rare, and it is occasionally suggested that it is a product of secret government experimentation.

COME ON, FIRST ONE TO INVERARY GETS A DEEP-FRIED MARS BAR

HARE

The animal called the hare is believed to bring bad luck, especially when it crosses one's path. This is somewhat difficult to avoid, however, as it is the fastest land animal in Scotland. It enjoys sports and will readily race other animals, including pet tortoises, despite an embarrassing incident many years ago.

DINNAE FRET YERSELF, HEN

HEN

The hen is a female feathered beast prized for its eggs. It is discerned from afar by its clucking noise. It is also common in Scotland for the hen to adopt a human form, whereupon the clucking may be more pronounced.

FABULOUS,
MAISIE,
FABULOUS

HIGHLAND COW

Tousle-haired and tan, the Highland cow is the poster-girl of the bovine field. Bullied and bull-imic, she is a warning to aspiring young heifers to beware when the grass appears greener.

HOODIE

The hoodie is a mysterious beast associated with fairies, and its appearance is often taken as a sign of foreboding. It is traditionally said to have the appearance of a crow, but in more recent times it is often reported that human features can be discerned.

HONESTLY, I CAN'T DO A THING WITH MY HAIR TODAY

KELPIE

The kelpie (or *each-uisge*), known as a 'nuggle' in Shetland and a 'tangie' in Orkney, is a water spirit, or *fuath*. Said to haunt streams and torrents, it has the appearance of a horse, but with webbed feet. It can shape-shift into a human form, though its disguise is usually given away by pond-weed in its hair.

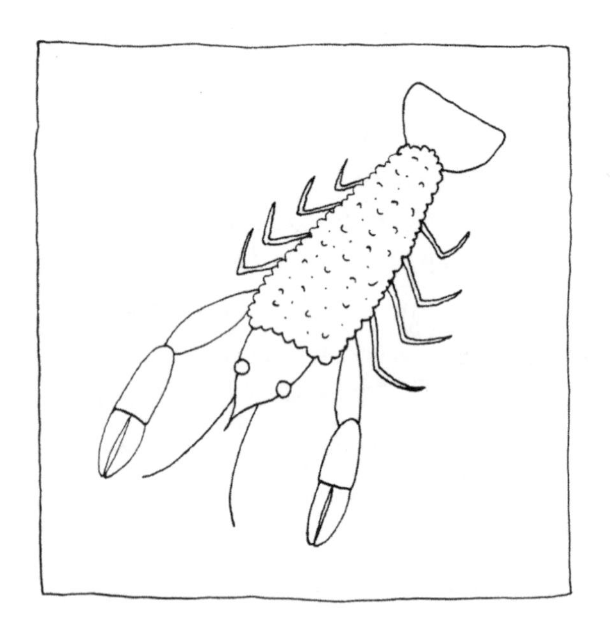

LANGOUSTINE

The langoustine, or scampi, is found in both the North Sea and the Atlantic. It is distinct from other crustaceans in that its outer crust is made of breadcrumbs rather than shell. Some believe them to be related to the Scotch egg, and it is true that they are occasionally found in the same oily waters.

AYE, YE
##*#*!

MIDGE

The midge is a wee biting beastie with a tenaciousness disproportionate to its size. Known for hunting in swarms, its favoured prey is the tourist. It is undeterred by rain, which is unfortunate, and will penetrate even the sturdiest of armour.

ROBERT, YOUR BLOOMIN' MOOSE IS LOOSE AGAIN

MOUSE

The mouse, not to be confused with the moose or certain whipped-cream desserts, is, according to Robert Burns, a wee 'tim'rous beastie', known for its ability to fit through small gaps and for making housekeepers jump onto chairs. Its favoured foods are cheese, peanuts and electrical wires. It is said to represent man's struggle with the world.

NESSIE

The Nessie, or Loch Ness monster, lives in the water. Nessies are mischievous beasts who will often collect and arrange objects on the surface to make people think they have seen them when they haven't.

ONESIE

The onesie is a beast with many diverse forms, but always with a human face. Adults are found mostly indoors, in front of televisions, but the young are often seen outdoors, usually in the company of humans.

OSPREY

The osprey is a sea bird known for its diving ability. Largely absent for nearly forty years, the osprey recolonised Scotland from Scandinavia in 1954 and can now be seen at sites throughout the country. They are protected by Operation Osprey, a volunteer organisation trained in birdwatching and military strategy.

OTTER

The otter is semi-aquatic and feeds mostly on fish. It is a good swimmer, its favourite style being the backstroke. The King Otter (Rìgh nan Dòbhran) has magic skin that protects it from harm, being vulnerable only on the white spot under its chin.

PINE MARTEN

The pine marten is so called because it likes hiding in pine trees. It has excellent sight, sense of smell and hearing, which it uses to find food and escape predators. It is a messy eater, but has a yellow bib to hide spills when eating honey.

PTARMIGAN

The ptarmigan is a small grouse that lives high in the mountains. Its genus name (*lagopus*) means 'hare foot', because it has feathers on its feet. It turns white in winter to better hide in the snow.

OOH, LOOK AT HIM WITH HIS FANCY BEAK

PUFFIN

The puffin is a sea bird much admired by humans, who find its appearance most appealing, despite the mouthfuls of fish. Puffins are found mostly in high rocky places by the sea, where they annoy other sea birds with the attention they draw.

RED DEER

The red deer is much revered. It is frightened of everything, which causes it to freeze at the slightest sound. Historically this has made it a favourite of hunters and painters.

HAVE YOU SEEN MY SCOTCH EGG?

RED SQUIRREL

The squirrel is so named because it squirrels things away for the winter. Its favourite foods are nuts and cones and picnic leftovers. The red squirrel is a shy wee ginger laddie that struggles to compete with the boldness of its grey relatives.

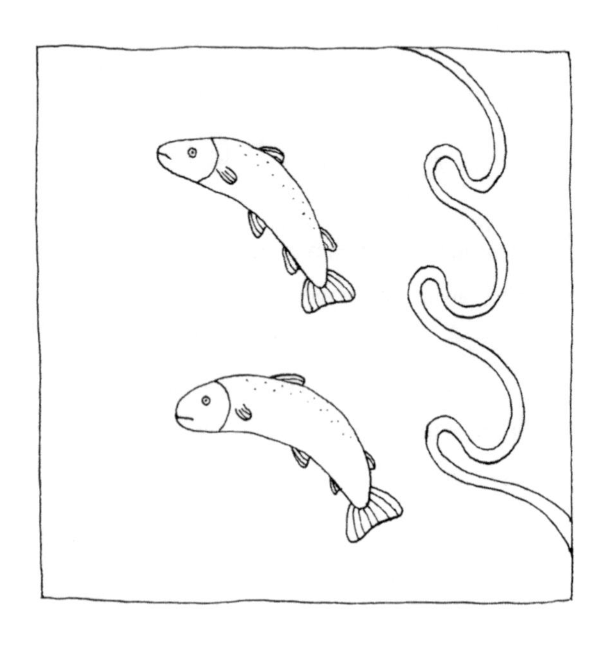

SALMON

The salmon is an aquatic beast much prized for its meat. Fishermen often report its size to be anything up to that of a small child, but history suggests their estimates are unreliable. It is famous for persistently trying to swim upstream to lay its eggs.

SASSENACH

The Sassenach, or Saxon, is a beast common throughout Scotland, despite numerous historical attempts to eradicate it. Mostly benign today, its numbers increase mysteriously in the holiday season. It can often be identified by walking shoes and a third eye on its chest.

SCOTCH EGG

The Scotch egg is produced by the Scotch chicken. It is often laid from a height, so it is protected by a layer of fatty meat and something resembling breadcrumbs. Despite this obvious biological explanation, there are claims that it is a human invention. It is so prized that few reach adulthood, so the eggs are now propagated artificially.

LEGENDARY TOURS

SCOTTISH WILDCAT

The Scottish wildcat lives mostly in woodlands, making only infrequent forays into civilisation for cat biscuits and litter trays. It is sometimes believed to be a much larger cat than it actually is, when it is given a name like 'the Galloway Puma'. This is reported mostly by tourists, but the reason for this is unclear.

THEY SAY IRN-BRU IS MADE FROM GIRDLES

SEAL

The seal is a playful animal that gives much pleasure with its aquatic antics. It is believed that larger seals are the children of kings under spells, or perhaps Norsemen making their way to Orkney or Shetland on the quest for Irn-Bru. These travelling shape-shifters are also called selkies.

SHEEP

The animal called the sheep is widespread throughout Scotland. It is famed for its woolly coat, which is often taken to make clothing. It feeds mostly on grass, which it consumes in great quantities before the onset of winter, sensing the severity of the season, but it has also been known to partake of the occasional trouser.

SHETLAND PONY

The Shetland pony is an ancient wee horse from a time when people were smaller. It is still common for them to be ridden, and because of their strength they are able to bear considerable weight. They are intelligent and wilful but, like other equines, are also susceptible to bribery, usually in the form of carrots or mint sweets.

WHAT DO YOU THINK OF THE HAGGIS FORMATION?

SPIDER

The spider is a small, many-legged beast. One once attempted to capture Robert the Bruce upon him entering its lair. Enamoured by the spider's boldness, the king offered it his friendship and they spent many hours in the cave discussing military tactics. The king went on to win the Battle of Bannockburn. The enemies of the spider are the shoe and the rolled-up newspaper.

SEE
YOU!

SPORRAN

The animal known as the sporran is a feisty wee beastie often captured and trained by men to protect the contents of kilts. Its sworn enemy is the handbag.

UNICORN

The unicorn is the national animal of Scotland, as seen on the medieval royal coat of arms. It is a creature of purity and grace, and so can only be captured by fair maidens, who often grow disinterested in them when they realise they still need mucking out.

DONALD, HAVE YOU SEEN HAMISH? HE'S BEEN DISMEMBERING TENNIS BALLS ON THE CHAISE LONGUE AGAIN

WEST HIGHLAND TERRIER

Of the diverse breeds of dogs, the West Highland white terrier, or 'westie', is a bristly, energetic beast, a master of deception, with its head often indistinguishable from its tail. It is not to be confused with household cleaning implements.

I'LL HUFF AND I'LL PUFF AND I'LL BLOW ON IT SO IT'S NOT TOO HOT

WOLF

The wolf is a fearsome beast known for its hunting. The Greeks call it *lykos*, which comes from their word for 'bites'. The true wolf has not been seen in Scotland for some time, but in Shetland there is a type of wolf-man called a 'wulver', who, legend says, is kindly and will often leave gifts such as fish to help poor families.

Tim Kirby is a British artist and illustrator. As well as private commissions, his work has also appeared in a large number of publications, and his clients include *Gramophone* magazine, the *New Statesman*, *Management Today* and the BBC. He is always courteous in his dealings with his fellow beasties. For more information, check out his website: www.timsgalleries.co.uk